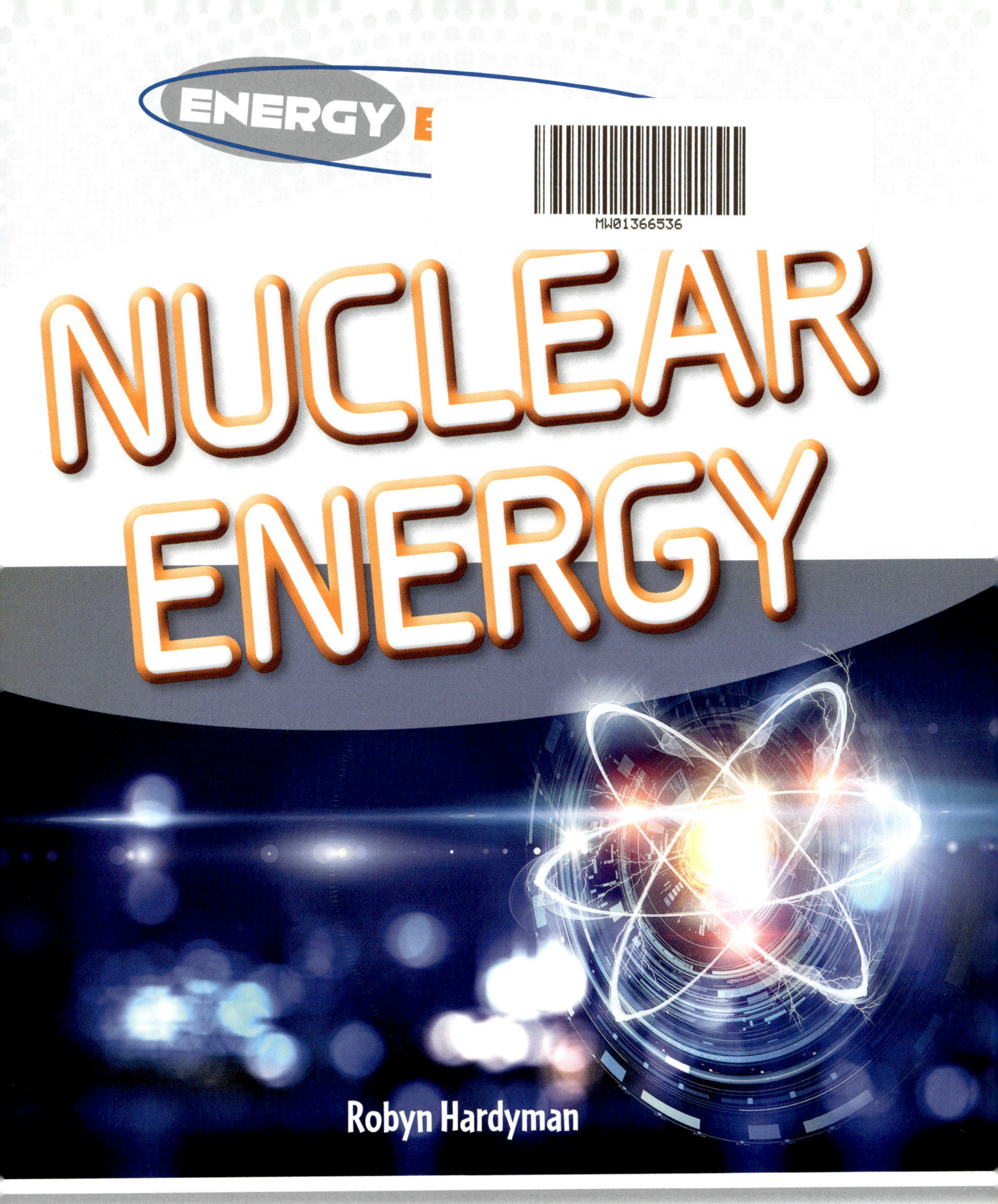

NUCLEAR ENERGY

Robyn Hardyman

CHERITON
CHILDREN'S BOOKS

Please visit our website, www.cheritonchildrensbooks.com to see more of our high-quality books.

First Edition

Published in 2022 by **Cheriton Children's Books**
PO Box 7258, Bridgnorth WV16 9ET, UK

Author: Robyn Hardyman
Designer: Paul Myerscough
Editor: Victoria Garrard
Proofreader: Wendy Scavuzzo
Picture Researcher: Rachel Blount
Consultant: David Hawksett, BSc

Picture credits: Cover: Shutterstock/Daniel Prudek; Inside: p1: Shutterstock/Sergey Nivens; p3: Shutterstock/Thorsten Schier; p4: Shutterstock/Sergey Nivens; p5: Shutterstock/Mandritoiu; p6: Shutterstock/Gritsalak Karalak; p7: Shutterstock/KREML; p8: Shutterstock/A_Dozmorov; p9: Shutterstock/Mark Higgins; p10: Shutterstock/Peter Sobolev; p11: Shutterstock/Peter Sobolev; p12: Flickr/Nuclear Regulatory Commission; p13: Flickr/Nuclear Regulatory Commission; p14: Shutterstock/Vladimir Mulder; p15: Shutterstock/Muph; p16-17: Shutterstock/Yevgeniy11; p18: Shutterstock/Smallcreative; p19: Shutterstock/Sergey Kamshylin; p20: U.S. Navy photo by Paul Farley; p21: NASA/JPLCaltech/MSSS; p22: Shutterstock/AlexKZ; p23: Wikimedia Commons/World Nuclear Association; p24: Shutterstock/Julius fekete; p25: Shutterstock/Arina P Habich; p26: Shutterstock/A. M. Teixeira; p27: Shutterstock/Matyas Rehak; p28: Shutterstock/Smallcreative; p29: Shutterstock/ Craig Hanson; p30: Shutterstock/Riekephotos; p31: Shutterstock/Sehenswerk; p32: Shutterstock/Thorsten Schier; p33: Shutterstock/Tatiana Zinchenko; p34: Wikimedia Commons/NuScale; p35: Flickr/Nuclear Regulatory Commission; p36: Wikimedia Commons/Energy.gov; p37: Wikimedia Commons/Nuclear Regulatory Commission from US; p38: Shutterstock/Markus Gann; p39: Shutterstock/Efman; p40: Shutterstock/Aerovista Luchtfotografie; p41: Wikimedia Commons/Oak Ridge National Laboratory; p42: Shutterstock/Serhat KINAY; p43: Shutterstock/Rangizzz; p44: Wikimedia Commons/Max-Planck-Institut für Plasmaphysik, Tino Schulz; p45: Shutterstock/Alones.

Printed in the United States of America

Contents

CHAPTER 1

What Is Nuclear Energy?

Nuclear energy uses the power that exists inside atoms—the tiny structures that everything in the universe is made of. A huge amount of heat energy is released when certain kinds of atoms are split. Nuclear energy does not release many harmful emissions **that contribute to** global warming**, but it does have some disadvantages.**

Nothing New

We have been using nuclear energy for many years, and it provides a sizeable amount of the world's electricity—about 11 percent. The first nuclear **power plants** were built in the 1950s, when scientists developed the technology to split atoms and release their energy. The number of nuclear power plants grew quickly, even though they are very expensive to build. Then, a series of accidents in nuclear power plants made people think carefully about whether nuclear power was the best solution for our growing energy needs. Over the past 10 years, the pressure has increased to "clean up" our energy act. We need to stop burning the **fossil fuels** that spew harmful gases into the **atmosphere**. People are now looking at how we can produce more nuclear energy, but safely.

When some types of atoms are split apart, they can release a huge amount of energy.

BIG Issues

Safety First

The biggest issue surrounding nuclear energy is safety. The process of splitting atoms produces radiation, which is a form of energy that travels through air. In high doses, radiation is very harmful to all living things. So if an accident occurs in a power plant and radiation leaks out, it can affect thousands of people. Also, the creation of nuclear energy produces waste. This can be harmful for thousands of years, so it has to be buried securely. These are the two main factors that have held back the development of nuclear energy over recent years.

The Good and the Bad

Nuclear energy is better for the **environment** than using fossil fuels to make electricity, and it is **sustainable**. It is also reliable. It can provide a constant supply of electricity, unlike other sources such as wind and **solar**. This makes it an attractive choice. However, it is not a **renewable** energy source, because there is a limited supply of the materials needed to create it. But we do have enough to last for about another 100 years.

Unlike with solar, wind, or **geothermal energy**, we cannot use nuclear power on a small scale. We cannot set up a mini plant to provide power just for our home or local community. To create nuclear energy, we must set up large power plants.

Nuclear power plants, like this one on the Hudson River in New York State, are often located beside water because they use the water for cooling.

How Does It Work?

Nuclear energy is created by splitting the atoms of a substance called uranium. The process is called nuclear fission. It must be done in very carefully controlled conditions because it creates radiation, which is extremely harmful if it escapes into the surrounding air, land, or water.

At the Heart of Atoms

Atoms are the tiny particles, or pieces, that everything in the universe is made of. At the core of an atom is the nucleus, which is where the term "nuclear" energy comes from. Each nucleus contains smaller particles called protons and neutrons. Even smaller particles, called electrons, circle around the outside of the nucleus, inside the atom.

In nuclear energy, the atoms of uranium are split. A neutron is fired at a uranium atom and strikes its nucleus, which is not stable. The nucleus splits into two lighter **nuclei**, plus some free neutrons. These free neutrons then crash into other uranium atoms, causing more splits. This is called a chain reaction because it can go on and on, with the free neutrons colliding with endless numbers of uranium atoms. This is what happens inside nuclear reactor power plants.

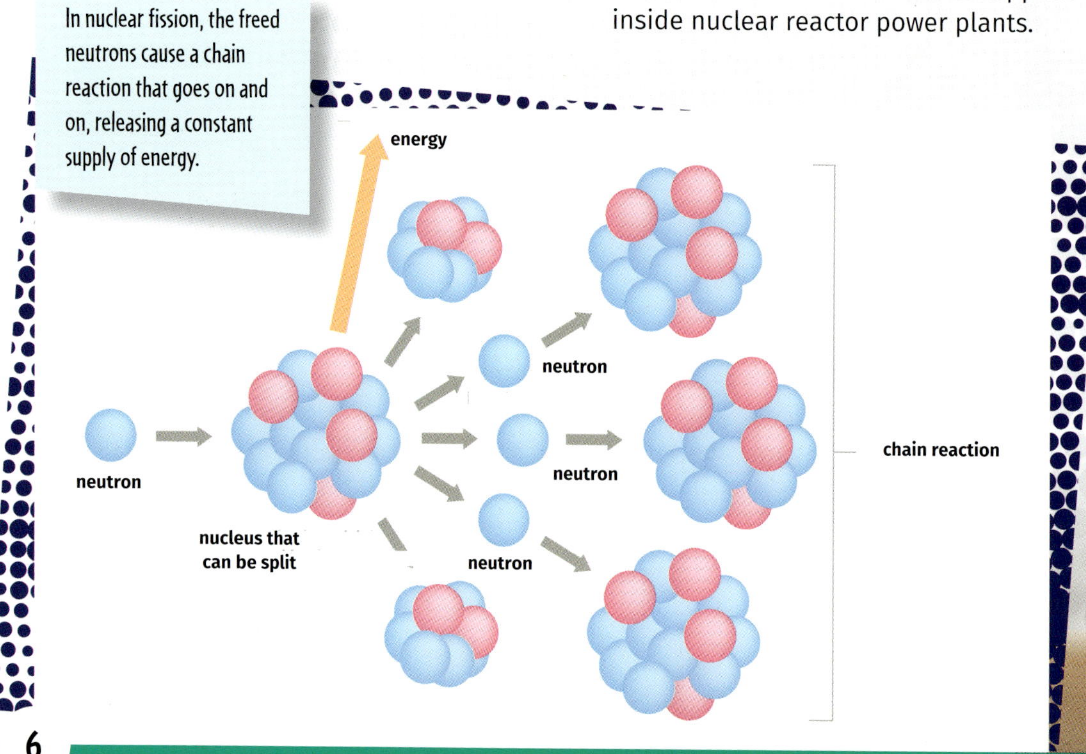

In nuclear fission, the freed neutrons cause a chain reaction that goes on and on, releasing a constant supply of energy.

What Happens in the Chain Reaction?

Some of the energy that is given off by the nuclear chain reaction is heat energy, and there is a lot of it. It is used inside the power plant to heat water and make steam. The steam powers the **turbines** in a **generator** to make electricity. The chain reaction must be controlled to create just the right amount of energy required.

The other energy created by the chain reaction is radiation, and that is the potentially harmful product. Radiation cannot be seen, but it can cause serious diseases such as cancer. Once it gets into the ground or the water supply, it remains there for many years, contaminating, or poisoning, crops and drinking water.

There are three types of rays in radiation: alpha, beta, and gamma rays. Alpha particles (protons and neutrons) can be stopped by a thin barrier, such as a sheet of paper. Beta particles (electrons) can be stopped by a sheet of **aluminum** 0.12 inches (3 mm) thick. Gamma rays, however, can only be stopped by a thick sheet of **lead** or a concrete shield more than 6 feet (1.8 m) thick.

Careful Work

Everyone who works with nuclear fuel and nuclear power wears special protective clothing. Experts monitor, or check, the level of radiation in the air around them very closely. The materials used are all handled from a distance, by remote control.

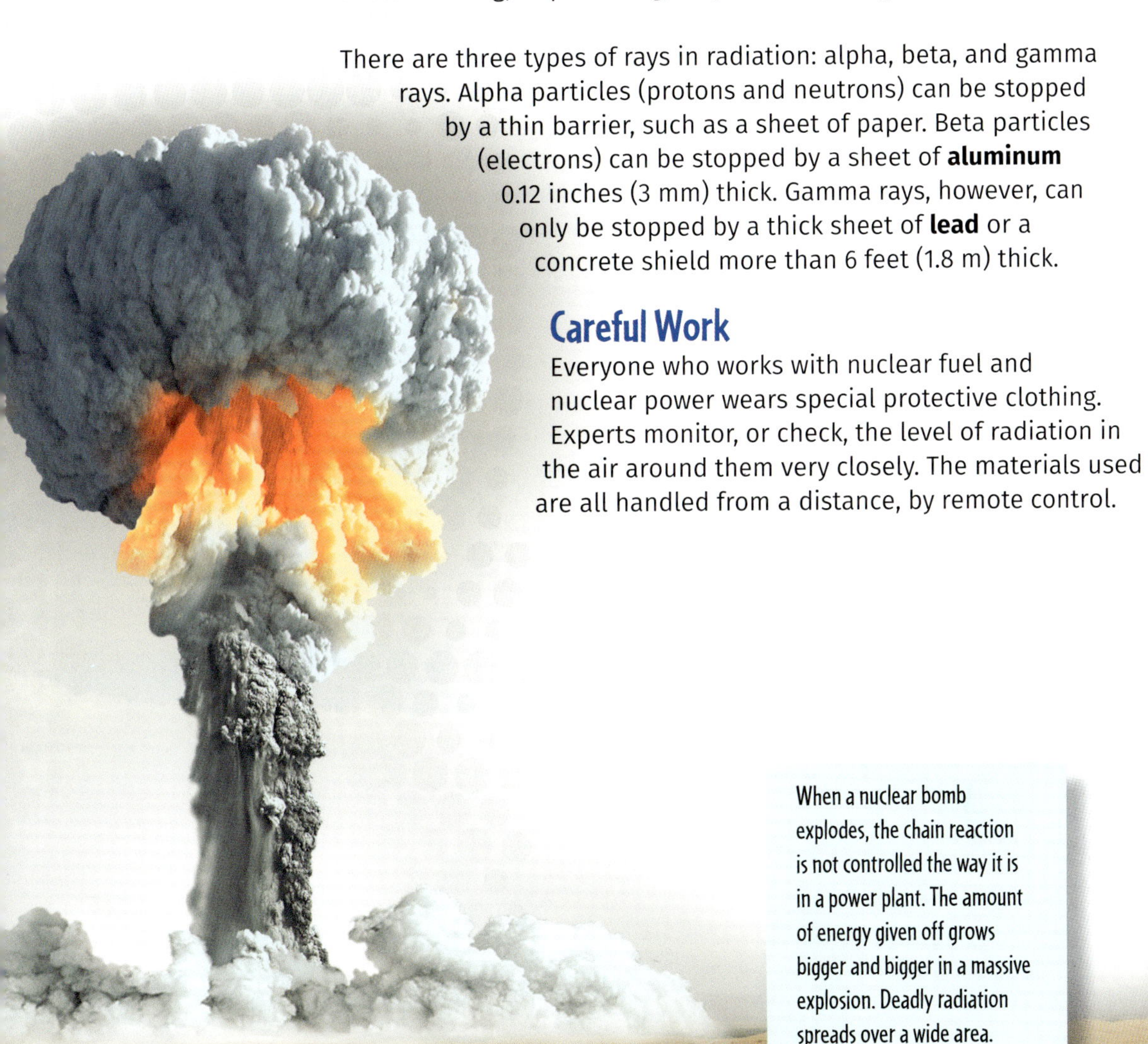

When a nuclear bomb explodes, the chain reaction is not controlled the way it is in a power plant. The amount of energy given off grows bigger and bigger in a massive explosion. Deadly radiation spreads over a wide area.

CHAPTER 2

The Fuel Cycle

Most nuclear power plants use uranium as their fuel. Uranium is found in many places across the world in a number of different forms. These forms are called isotopes. Only some of these isotopes can be used for nuclear power, and this makes the supply of nuclear fuel more limited.

The Best Uranium

Uranium is a silvery-white metal. It is found in rocks, called ores, such as granite, sandstone, and pitchblende. It has to be extracted from these rocks before it can be used. The uranium isotope that works best for nuclear fission is called uranium-235, or U-235. It is so-named because it has 92 protons and 143 neutrons in its nucleus. These add up to 235. It is very unstable, and it releases a lot of energy when it breaks apart. Although uranium is about 100 times more common than silver, U-235 is quite rare.

This rock contains uranium-235 that must be extracted before it can be used.

Getting to the Uranium

About half of the world's uranium comes from 10 mines in six countries: Canada, Australia, Niger, Kazakhstan, Russia, and Namibia. The earliest mines were huge, open-pit mines on the surface. The layer of soil at the surface was stripped away with giant scrapers. The ore that was revealed was dug out of the ground, loaded onto huge trucks, and taken away to be **processed**.

The industry has developed a new technique, however, now used by about half the world's mines. This is called in situ leaching, which is better for the environment. In this process, the uranium is mined without major ground disturbance. **Groundwater**, with a lot of oxygen injected into it, is pushed through the uranium ore underground, to extract uranium. The water containing the uranium is then pumped to the surface. The uranium oxide is extracted from it and dried. It is now called "yellowcake" due to its color, and it is still only slightly **radioactive**.

BIG Issues

Finding Uranium

Finding where uranium lies underground can be done in several ways, such as exploratory drilling into the ground or by taking **samples** of the soil and groundwater in likely places. A more recent, and more high-tech, method is to study the land from the air with instruments that can measure the amount of gamma-ray radiation being given off by the ground. This builds into a picture of where uranium may be present.

This uranium mine in Australia is now closed. It mined the rock in the old-fashioned way—by scraping the ground away with heavy machinery.

How Nuclear Fuel Is Made

Almost all nuclear power plants need uranium with a higher amount of U-235 in it than usually occurs naturally. This is called "enriched" uranium. Enriching uranium is very expensive and uses a lot of energy.

Enriching the Uranium

The uranium has to be enriched by increasing the amount of U-235 in it from about 1 percent to 3 to 5 percent. This has to be done with the uranium in gas form, so once it has left the mine, it goes to a plant where it is processed into a gas. The gas is separated into two streams: one with the enriched uranium and the other with the remainder. Only about 15 percent of the gas becomes the useful stream. The other stream is called **depleted** uranium, and it cannot be used to make nuclear power.

The industry has developed a newer method to separate the U-235. This method spins the uranium at top speed in thousands of vertical tubes. An even newer **innovation** is looking at using **laser beams** to do that.

This processed uranium fuel is now ready to be packed into tubes and used in a nuclear reactor.

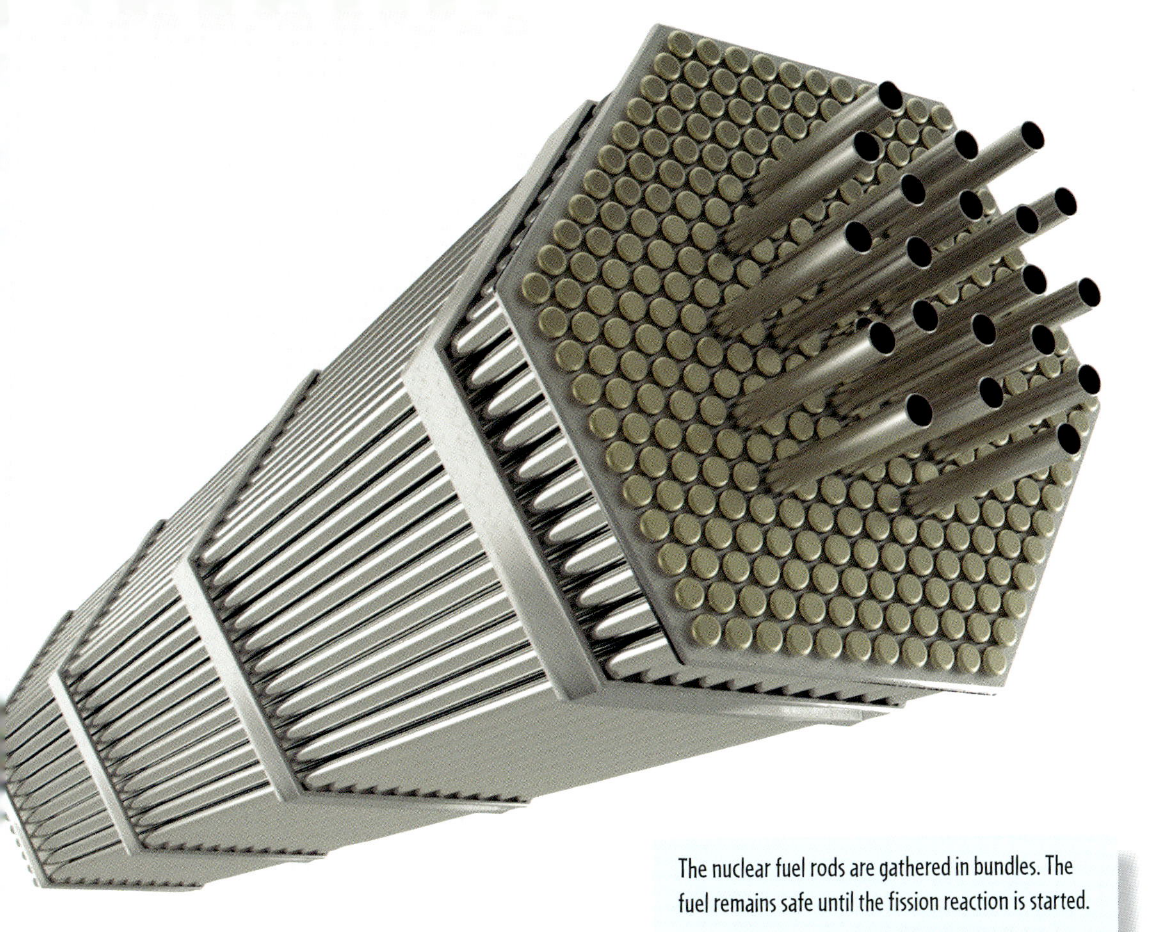

The nuclear fuel rods are gathered in bundles. The fuel remains safe until the fission reaction is started.

Making Nuclear Fuel

The enriched uranium is now ready to be made into nuclear fuel in a processing plant. It takes much less nuclear fuel than coal to produce electricity. For example, 30 tons (27 metric tons) of nuclear fuel can produce the same amount of electricity as 2.8 million tons (2.5 million metric tons) of coal. This is a major advantage of nuclear energy. At the processing plant, the U-235 is converted to a powder, which is now black. The powder is then formed into small pellets that are heated until they form a hard material. These are packed into stainless steel tubes about 13 feet (4 m) long to form fuel rods. Finally, the rods are grouped together into bundles.

Trucks then carry the bundles to the nuclear power plants. There, they are stored until they are needed. The number of fuel rods needed depends on the design of the nuclear power plant. Some plants have more bundles, each with fewer rods, while others need fewer bundles, but each contains more fuel rods. Most power plants need to replace about one-third of all their rods every 12 to 24 months.

The fuel may be stored for some time, but it is still only mildly radioactive because the fission reaction has not begun. Any radiation that is there is safely contained within the metal tubes of the bundle.

Getting Rid of Waste

Once the fuel has been used to do its job of making electricity, it is highly radioactive. It must be handled extremely carefully afterward. Although nuclear reactors can produce relatively cheap energy, it takes a lot of energy and money to clean up the waste they create. This is the end of the fuel cycle.

Keeping Things Cool

The bundles of used fuel rods are very radioactive and incredibly hot. To be stored safely, they must first be cooled. Amazingly, just a few feet of water is enough to provide a safe shield that prevents the radiation from affecting anyone or anything nearby. The fuel rods are removed from the bundles underwater, then kept underwater in huge pools. An hour or two is not enough to remove their heat, though, because the fission is still going on. The rods must spend about five years in the water before they are cool enough, and safe enough, to be moved. In some cases, the rods stay in the water for up to 50 years. In others, they are removed and transferred to huge, dry concrete or steel containers. These are cooled with air to keep the heat from building up.

This pool is holding fuel rods at a nuclear power plant in California.

Figuring Out the Problem

The nuclear industry has not yet found a permanent way to store its waste. The pools and the dry storage containers are temporary solutions until people figure out an answer to the problem. Experts think that the waste will have to go to specially engineered facilities deep underground, in places that cannot be affected by earthquakes.

Turned into New Fuel Rods

In a few countries, the fuel rods are taken to a plant after they have spent long enough in water. There, they are **dissolved** in **acid**, and the uranium in them is removed. It can then be made into new fuel rods. This process also creates another substance called plutonium. Plutonium can be used as a fuel for a particular kind of nuclear power reactor called a fast neutron reactor.

Low-level waste from the power plant, such as containers and clothing, is buried in pits underground to keep it safe.

BIG Issues

Other Waste

It is not just fuel rods that are radioactive. A lot of other waste from power plants is radioactive, such as tools, work clothing, and steel parts from within the reactor. These are not nearly as radioactive as the fuel rods, but they must still be disposed of safely. There are regulations, or rules, to control this, so the waste never comes into contact with the outside environment.

CHAPTER 3

Inside a Reactor

What actually happens inside a reactor, at the heart of a nuclear power plant? How is the nuclear fission managed, so that the chain reaction does not get out of control? Let's take a look at the different types of nuclear reactors used in power plants, and how they work.

Controlled by Water

The most common type of nuclear power reactor is the pressurized water reactor. In its core, or center, the fuel rods are used to heat highly pressurized water, which produces steam to power the electricity generator. Water is the moderator, or controller, that keeps the fission process in check. It slows down the neutrons that are produced by the fission. A pressurized water reactor uses 120 to 190 fuel bundles, each with 180 to 260 fuel rods. The rods last about a year before they need to be replaced. These reactors are often built close to rivers or the ocean, so that they have a constant supply of water.

Operations at a nuclear power plant are controlled by experts in a central control room.

These steam turbines in the turbine hall of a nuclear power plant are huge machines. They use steam to generate electricity.

Keeping Cool with Gas

Another type of reactor uses the gases **carbon dioxide** or **helium** as **coolants** instead of water. The moderator is also not water. It is graphite, which is a form of carbon. Reactors in the United Kingdom (UK) and France use this kind of design.

Safety Worries

Everything in a nuclear power plant is carefully designed to protect the outside world from any danger. The core of the reactor sits inside a steel vessel, with the inside at high pressure. This keeps the water around it in liquid form, even though the temperature reaches more than 608 degrees Fahrenheit (320 °C). The steam to power the generator is formed in separate vessels. After use, it is condensed, or squashed together, and the water is **recycled** through the core.

The building is also protected by concrete walls more than 3 feet (1 m) thick. This is in case of an explosion, but also to contain potential leaks. Many nuclear power plants have a domed (curved) roof on the containment building, which is there to contain steam and radiation in case they leak out of the reactor. The biggest risk is that the supply of cooling water is disrupted. In these instances, the reactor will shut itself down, and an emergency core-cooling system will start flooding it with a separate supply of cold water.

Working Around the Clock

Nuclear power plants are very expensive to build. But once they are running, they are relatively cheap because all the fuel is already in place. It makes sense, therefore, for them to operate for as many hours of the day as possible. About 20 years ago, they ran for 60 to 70 percent of the time. But with improvements in **efficiency** in recent years they can now run for about 90 percent of the time.

New Designs

The nuclear power industry is changing. New techniques to make plants more efficient **and safer are being introduced all the time. Although nuclear power has been around for more than 60 years, the latest reactors are very different from the reactors used in the earliest plants.**

Fast Breeder Reactors

A different type of nuclear reactor has the amazing ability to create new fuel faster than it uses it. This is called a fast breeder reactor. It is "fast" because it does not slow down the neutrons during the fission process. Fast breeder reactors were developed when it seemed as though the supply of uranium was going to run out in a few years. Since that time, however, we have come up with better ways of finding and processing uranium. This has resulted in more nuclear fuel that is cheaper to buy. For these reasons, there is less need for fast breeder reactors today.

Fast Neutron Reactors

The latest development in reactor design is the fast neutron reactor. This is very different from the pressurized water reactor. Fast neutron reactors use plutonium as well as uranium as fuel. They can also use a second isotope of uranium—U-238—as well as U-235. They do not always produce more fuel than they use, but they are extremely efficient at making energy with the fuel they have. In fact, they can get as much as 60 times more energy from these fuels as from other reactors.

Fast Neutron Future

Another major advantage is that fast neutron reactors can burn a part of the fuel that other, ordinary reactors cannot, and which they have to get rid of as radioactive waste.

A number of fast neutron reactor plants are being tried out around the world. The World Nuclear Association (WNA) expects there to be a rise in fast neutron power plant design and use around the world in the future.

BIG Issues
Keeping Cool

Even in fast reactors, for safety reasons, it is very important to keep the fuel rods in the core cool. Instead of water, the coolant is a liquid metal, usually sodium. It does not slow the fission, but it keeps the temperature under control.

This fast neutron reactor was built in Russia to try out the technology, but it is no longer in operation.

Is It Safe?

Nuclear energy has some big advantages over fossil fuels. The level of harmful emissions produced is far lower, and it is an efficient way to make electricity from fuel. We have been using it for more than 60 years, so why is nuclear energy not providing the power for all our needs? The main reason is that some people do not trust the safety of nuclear power or the radioactive waste that it produces.

When Disaster Strikes

There have been some terrible nuclear disasters in the history of the industry. In 1979, at the Three Mile Island plant in Pennsylvania, there was almost a "meltdown." This means that the cooling system failed, and the fuel rods nearly burned through the reactor into the ground. The accident was brought under control, but many newly planned power plants in the United States were canceled as a result.

The accident at Chernobyl in the Ukraine in 1986 was much worse. The rods overheated and broke up inside the reactor. As they touched the coolant water, there was a giant explosion of steam that blew the roof off the reactor. Radioactive material and gases were thrown high into the air. The radiation was carried by the wind over an enormous area across Europe. It took two weeks just to put out the fire and stop the fission reaction. Many people lost their lives, and thousands more suffered from the effects of the radiation for years afterward.

In 2011, the reactors at Fukushima in Japan suffered three core meltdowns after the plant was hit by a giant wave called a **tsunami**. More than 450,000 people were forced to leave their homes.

Safety is the nuclear energy industry's highest priority, because radiation is so harmful to living things and the land for a very long time. This photograph is of the now-abandoned Lost City of Chernobyl, where one of the worst disasters took place.

Built for Safety

We have seen how nuclear plants are built to cope with huge pressures, and since terrible accidents took place, the regulations surrounding nuclear power have tightened up a lot. Every detail of the building of plants is controlled, and there must be backup cooling systems in place to keep the fuel cool if the power or the water supply fails. The regulations are strict about maintenance, too, and all the workers are highly trained and skilled.

BIG Issues

Waste Worries

The biggest question is: What do we do with the waste? Nuclear power plants in the United States produce around 2,204 tons (2,000 metric tons) of waste each year. That waste remains radioactive for thousands of years. It is risky to transport it by truck over public roads, and there is still no place to store it in the long run. Most plants, therefore, store it on site in huge concrete and steel casks, which must be closely guarded and maintained. A plan to build a huge waste dump deep under Yucca Mountain in Arizona has been halted because so many people were against the idea.

A Lot of Different Uses

The energy that is contained inside the nucleus of uranium and other materials can be used for purposes other than providing electricity for the grid. Let's take a look at some of them.

This nuclear-powered attack submarine is operated by the US Navy. It can stay out at sea for a long time without refueling.

Ships and Subs

Nuclear power is particularly suitable as a source of power for vessels that need to be at sea for a long period of time without being able to refuel. These can include the navy's aircraft carriers or icebreakers in the polar regions. The most common use for nuclear power at sea, though, is for submarines.

Each nuclear submarine has its own small pressurized water reactor on board. They work in the same way as large reactors, but the heat makes steam that powers the propeller that pushes the submarine through the water. The reactor can run for up to seven years without refueling, so the submarines can stay underwater for months at a time.

Nuclear in Space

There is another location where refueling can be tricky: space. A special kind of reactor called a radioisotope thermoelectric generator (RTG) is used in space missions. There is no chain reaction in these reactors. Instead, the heat generated by the decay, or breakdown, of a radioactive source, such as plutonium, is used to generate electricity. These radioactive materials naturally decay over time, becoming less radioactive and releasing heat as they do so. The Voyager **space probes**, the Cassini mission to Saturn, and the Galileo mission to Jupiter were all powered by RTGs. They all traveled for many years, and the two Voyager probes, which were launched in 1977, are still in space today. Nuclear power has supplied their electricity needs during their journey to beyond the edge of our **solar system**. The 2006 New Horizons mission to Pluto is also powered this way.

Exploring Mars

Nuclear power can also be used to power craft that land on other planets. The rovers that landed on Mars, Spirit and Opportunity, used **solar panels** for electricity. The latest Mars rover, Curiosity, is much bigger, and it uses RTGs for both heat and electricity because solar panels would not be able to supply it with enough power.

The Idaho National Laboratory's Center for Space Nuclear Research (CSNR) is working with the National Aeronautics and Space Administration (NASA) to develop an RTG-powered hopper vehicle for Mars exploration. This would have a core of nuclear fuel that would use the carbon dioxide in the air on Mars to create a mini explosion. This would allow it to "hop" 9.3 miles (15 km) at a time. This would give it access to areas of Mars that traditional rovers cannot reach.

The latest rover on Mars, Curiosity, uses nuclear energy as its source of power.

CHAPTER 4

Where in the World?

Nuclear energy provides about 11 percent of the world's electricity today. This is generated by about 450 nuclear reactors in 30 countries.

A Big World Player

Nuclear energy is the world's second-largest source of low-carbon power, providing about 30 percent of the total. Even countries that do not have their own nuclear power plants are benefiting from nuclear energy, because they import, or buy in, electricity from countries that do. Italy and Denmark, for example, get about 10 percent of their electricity from imported nuclear power.

In 2017, 13 countries produced at least one-quarter of their electricity from nuclear energy. France has always been a world leader in nuclear energy, and produces around three-quarters of its electricity this way today. Hungary, Slovakia, and Ukraine get more than half of their power from nuclear energy, while Belgium, Sweden, Slovenia, Bulgaria, Switzerland, Finland, the Czech Republic, and South Korea get around one-third. In the United States, the UK, Spain, Romania, and Russia, about one-fifth of electricity comes from nuclear energy. The United States generates more nuclear energy than any other country, but that is a lower proportion of its total electricity generation.

This mine is extracting uranium ore from the ground in Kazakhstan in Central Asia. Kazakhstan is one of the world's leading producers of uranium.

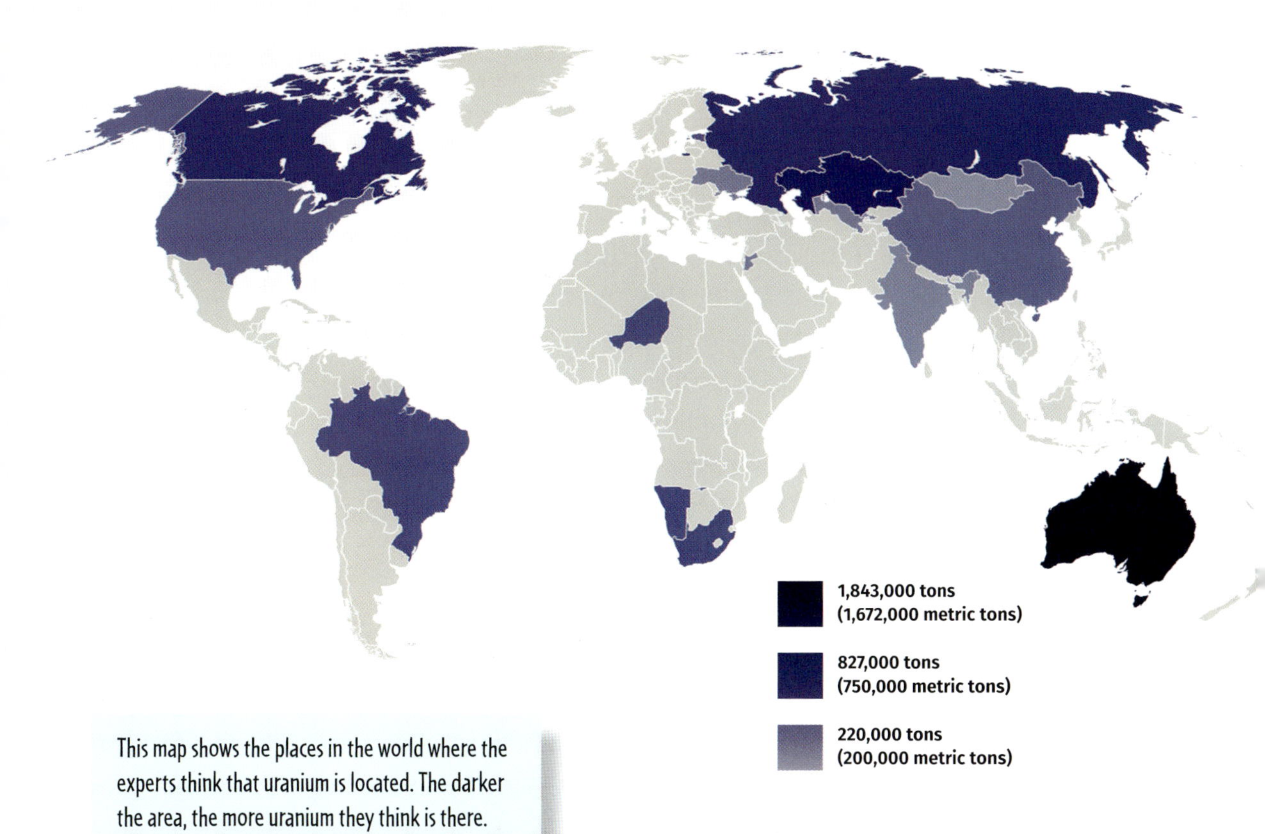

This map shows the places in the world where the experts think that uranium is located. The darker the area, the more uranium they think is there.

The Money Problem

Nuclear power is expensive to develop. The costs of mining the uranium, processing it, then building a plant to strict safety standards, have meant that poorer countries have not developed nuclear technology to the same extent as wealthier nations. The world's leading producer of uranium is Kazakhstan, while Canada and Australia are also significant producers. There are supplies that have been identified, but not yet extracted, in Asia, Africa, and Europe. Around 65 new nuclear reactors are under construction around the world, mostly in China, India, Russia, and South Korea. Those countries have big plans to grow their nuclear power programs in the future.

BIG Issues

Enough to Last?

If we want to grow our nuclear power industry, we need to know we have a reliable supply of fuel in the form of uranium. The International Atomic Energy Agency (IAEA) made an assessment of the uranium resources across the world. It showed that the total amount we have identified in the ground has risen by around 70 percent over the last 10 years. It looks as though we have enough to last for another 100 years, if we continue using it at the same rate as we do now.

Nuclear Energy in the United States

The United States generates more electricity from nuclear power plants than any other country in the world. The country has around 90 nuclear reactors in operation in 30 states, operated by 26 different power companies. These plants are efficient, too, operating at around 90 percent efficiency. That is a huge improvement from the 1970s, when the plants were only producing power around 50 percent of the time. This is an industry that is getting better and better at what it does.

A Long History

The first nuclear reactor in the United States began operating in Pennsylvania in 1957. That was three years after the Soviet Union began operating the world's first nuclear reactor. Many more were built in the years that followed, but growth slowed in the 1990s. Nuclear power plants in the United States are mainly in the eastern and central states, though there are some in California, Arizona, and Washington. Almost all of their fuel is imported from overseas, with most coming from Canada, Australia, and Russia.

This nuclear power plant is in San Onofre, California, on the Pacific coast.

The nuclear power industry in the United States supplies electricity for one in five homes and businesses.

Changes in the Industry

The number of reactors has fallen slightly in the last 25 years, however. This is partly because some of the older plants have reached the end of their expected life. Most of them were built from 1967 to 1996. It is also because of the increase in the supply of cheap natural gas as an alternative fuel. This has made some nuclear power plants in the United States less **competitive** in the cost of the energy they supply.

The newest reactor in the United States entered service in 2016. This was the Watts Bar reactor in Tennessee, the first new reactor to open in the country since 1996.

Despite some reactors closing, the amount of electricity generated is still about the same. This is because new technology is making them more efficient.

Brand New Power

In Georgia, a new, huge nuclear power plant has been built—the Vogtle Units 3 and 4. During building, construction costs rose over and over. But, despite this, the energy companies behind the massive project decided to continue with it. The US Department of Energy (DOE) has said that it hopes this will be the start of a revival in the nuclear industry in the country.

Nuclear Energy in Latin America

In the countries of Central and South America, as in the rest of the world, there is an urgent need to find new, cleaner ways to produce electricity. These countries have traditionally been dependent on oil and gas for their energy. But now they are looking at alternative energy sources to help in the battle to limit global warming and climate change**.**

Off to a Slow Start

At the moment, there is little nuclear power in Latin America. There are just seven nuclear reactors in operation, producing little more than 2 percent of all the energy consumed. Three reactors are in Argentina, creating 5 percent of its electricity. The two reactors in Brazil produce 3 percent of its needs, and the two in Mexico make around 6 percent.

There are plans, however, for some of these countries to develop their nuclear energy capacity over the next few years. In Argentina, for example, an agreement was signed with China to develop two nuclear power plants at Atucha in Buenos Aires. In 2014, another agreement was signed with Russia to explore the possibility of a fifth unit at another site.

This reactor in Rio de Janeiro is one of only two in Brazil.

This is the model for a large new nuclear power plant in Cuba, but the project was put on hold, and nuclear power is now banned in the country.

Since 2014, however, work on bringing both of these projects into operation has been **postponed**. The main reasons were financial and the lack of support from the government.

An Uncertain Future

Brazil is also looking to a more nuclear future. In 2006, plans were announced for eight new plants to be opened by 2030. Since then, however, political difficulties in the country have slowed the projects. In Bolivia, cooperation agreements were signed with France and Russia in 2014. The Russian project was for a nuclear research center to help the country develop its nuclear industry. This was a sign of a country wanting to join the big players in the global nuclear community.

BIG Issues

Money, Money, Money

One of the biggest barriers to developing nuclear power plants in a country is cost. Governments need to offer financial help to the developers, to get them through the initial high start-up costs. Even then, nuclear energy may not be able to compete with the cheaper cost of gas-generated electricity.

Uruguay, Chile, and Cuba also explored the possibilities of using nuclear power, but after the accident at the Fukushima power plant in Japan in 2011, they stopped their plans. In fact, in Uruguay and Cuba, nuclear energy is banned. Mexico has also decided to stop the construction of 10 new nuclear plants, because the price of gas has fallen so much, and they intend to use it instead.

Nuclear Energy in Asia

In 2011, one country in the Far East—Japan—experienced the world's worst nuclear power disaster of recent years. Another country, China, is developing nuclear power faster than anyone else in the world. Nuclear power has a long history there, and it looks as though it has a bright future, too.

Nuclear Power in Japan

In 2011, a powerful earthquake off the coast of Japan caused a massive tsunami to crash on a long stretch of land along the coast. Thousands of people lost their lives in the disaster. The nuclear reactors at Fukushima registered the earthquake on their systems, and they shut down as they were supposed to do. The backup cooling systems started working, but then the tsunami hit the plant. All systems were swamped, and the cores in three of the reactors began to overheat. All three suffered meltdowns, the most severe kind of nuclear energy accident.

As a result of this disaster, all the country's reactors were shut down. They were not reopened for four years, leaving Japan with no nuclear power at all. The restart began slowly in 2015. Plants had to show that they had put in place even stricter safety controls to cope with the strongest earthquakes and tsunamis. Since then, 20 of the country's 37 reactors have been shut down permanently. By early 2019, only nine of the rest had been brought back into operation. Before the accident, Japan got one-third of its electricity from nuclear energy; today, it is closer to 4 percent. Still the government plans to increase that to around 20 percent by 2030.

The tsunami that hit the west coast of Japan in April 2011 caused a massive amount of damage and cost around 20,000 lives.

South Korea's Successes

South Korea, on the other hand, is a big nuclear energy success story. This small country has 23 reactors and, together, they generate almost one-third of the country's electricity. Four more reactors are being built there, and another two are planned. This is a country at the cutting-edge of reactor design, too, with a lot of research happening in this field.

Leading the Way

China leads Asia in nuclear power with 46 reactors, yet they still only provide 4 percent of its electricity. This is set to change because China is committed to cleaning up its highly polluted air. This pollution has come from coal-fired power plants. In early 2019, 11 new nuclear reactors were being built, and the country was working on some all-new designs for reactors. Even more **ambitiously**, it has plans to export, or sell overseas, its newly designed Hualong One pressurized water reactor. Finally, China is taking the long view. By the middle of this century, it plans to have mainly fast neutron reactors to replace the pressurized water reactors.

This nuclear power plant at Lianyungang in China is part of the country's big plans for nuclear energy.

Nuclear Energy in Europe

The countries of Europe have some very different views on nuclear energy as the source of their electricity. While France has been a nuclear pioneer and leader, for example, Germany is scaling back on its nuclear reactors and plans to have closed them all over the next 10 years. Today, nuclear power provides more than one-quarter of all European electricity, which is more than half of its clean electricity.

France, Making Changes

France has 58 nuclear reactors, but many of them are old and coming to the end of their useful life. The country is not rushing to close them, however, since they make an important contribution to its clean energy targets. It is developing other renewable sources and investing in new nuclear technology, while allowing the existing reactors to help keep the lights on.

Other Europeans

Germany made the big decision to close half of its nuclear reactors and shift to natural gas for its power plants. But this created a considerable rise in its emissions of carbon dioxide, and other countries are not willing to do the same just yet.

This nuclear fuel reprocessing plant is in France, where nuclear energy has been promoted for many years.

The power delivery systems across Europe are not designed to take the irregular supply of electricity that can come from renewable sources, but nuclear energy provides a steady supply.

Sweden, for example, is closing the old reactors that must close, but improving and keeping others where that can be done. The UK has some of the oldest reactors in Europe, but there are plans for two new ones to replace them, if a solution to funding them can be found.

Working Together

In 2013, 12 European countries joined to promote the role of nuclear energy. They are the UK, Finland, France, Netherlands, Spain, Bulgaria, Czech Republic, Slovakia, Hungary, Lithuania, Poland, and Romania. The countries of Eastern Europe want to reduce the amount of gas they import from Russia, so they are eager supporters of nuclear energy, too.

BIG Issues

Energy All the Time

One of the big advantages of nuclear power in Europe is that it can provide a regular and constant supply of electricity. The power lines across Europe are not designed to cope with the variable power supply produced by some renewable sources, such as solar and wind power. Since these supply networks often cross borders between countries and serve more than one country, it is very important that they are up to the job. The European Union (EU) recognizes that nuclear power must remain part of its energy picture while this is the case.

CHAPTER 5

New Ideas

There are many new technologies under development in the nuclear energy industry. They offer a constant stream of new ideas and techniques, which aim to lower costs and make plants more efficient, flexible, and safe.

It is hoped that advances in the design of nuclear power plants, like this one, will make the industry safer and more efficient in the future.

Faster, Better, and Cheaper

Developers are creating simpler designs for reactors, so that some of the basic parts can be built off-site in the factory and brought to the power plant more complete. This makes building the reactor faster and cheaper. The reactor size can also be more varied. This allows developers to choose a plant size suited to the needs, and **budget** of the users. This more-flexible approach is ideal for smaller companies and for places in the countryside, where demand for electricity is lower.

This is the steam generator of Novovoronezh nuclear power plant in Russia. A new plant, using the latest innovations, is being built on the site.

Gearing Up the Grid

Today, advanced reactors are also more able to adjust their energy output to match demand. This has become very important because the new renewable energy sources often deliver highly variable amounts of electricity through the day and night. The grid finds this difficult to deal with, so if nuclear-generated electricity can be adjusted to make up for spikes or drops in the power supply, that would even out the supply and make the grid work most effectively.

Safety First

Many important lessons have been learned from the Fukushima nuclear accident in Japan in 2011. The industry has developed a series of measures called FLEX. These protect nuclear power plants against terrible natural events happening simultaneously at more than one reactor. More backup power and water supplies are being put in place, along with many other new safety measures.

Useful Waste

Nuclear waste remains radioactive for a very long time, but scientists have found a way to make use of that energy. This is called betavoltaics. The beta particles, or electrons, given out by the radioactive waste are collected in a simple battery and used to generate an electric current. The amount of electricity is low, but it can last for many years, while the material remains radioactive. This makes betavoltaic devices perfect for **remote** and long-term use, such as in spacecraft. Only a small amount of radioactive material is needed, so the device can be very small.

The steam from advanced reactors can also be put to other uses. They work at higher temperatures than older reactors, so there is more steam available to be used by other industries nearby. For example, the steam can provide heat for processes used in the making of **fertilizer** for farming.

Small but Smart

One new design in nuclear reactors is promising to make the industry more flexible and cost effective. This is the small modular reactor, or SMR.

Cheaper and More Flexible

One of the advantages of renewable energy sources, such as solar and wind power, is that they can be scaled down and used to serve just a community or even a single building. Nuclear energy plants, on the other hand, have been growing bigger and bigger over the last few decades. This makes them very complex and extremely expensive. Although nuclear power generation can never get down to the single-household level, innovations in reactor design are now making it possible for smaller plants to be built.

Many Positive Points

SMRs are manufactured at a plant and taken to the nuclear site to be fully constructed. They can be used singly in places where less electricity is needed, or several can be used on one site, so the amount of electricity the site produces can be varied more easily to meet any demand. Another advantage is that these smaller reactors can often fit into locations where a standard nuclear power plant could not, such as the sites of old coal-fired power plants.

SMRs are more flexible than the old, huge reactors. They are built off-site and can be installed at smaller locations.

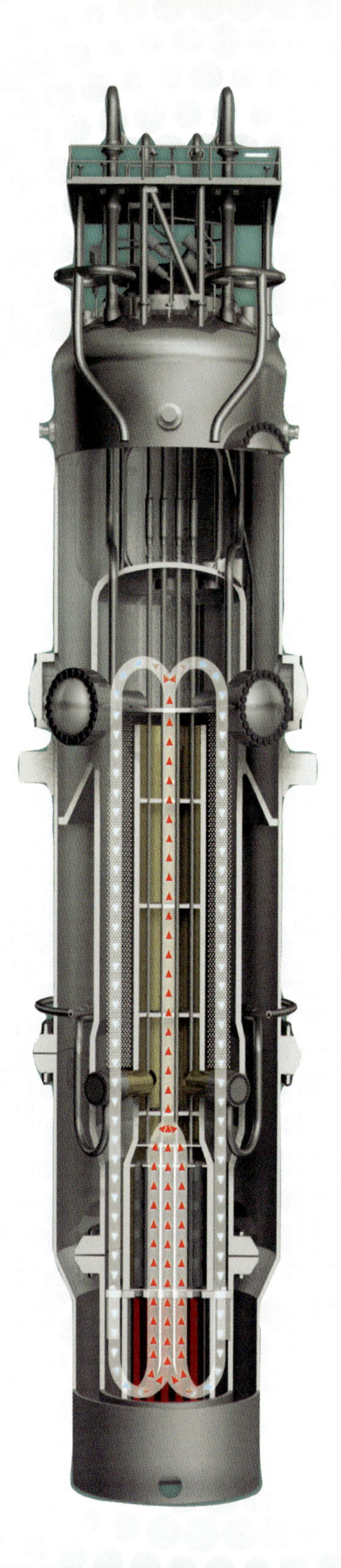

Senior figures in the nuclear industry are meeting to discuss the introduction of SMRs at an annual industry conference.

Making It Happen

There are several different designs proposed for SMRs. Some are simply scaled-down versions of existing nuclear reactor designs; others are entirely new. They include fast neutron reactors as well as traditional pressurized water reactors. The IAEA estimates that there could be at least 40 SMRs in operation around the world by 2030. There is research being done in several countries and, so far, China has progressed the furthest in getting these new reactors going. In the United States, the Tennessee Valley Authority has plans to build an SMR at the Clinch River site near Oak Ridge, Tennessee. This is the first plan in the country for an SMR.

Safety Advantages

These smaller reactors may also have safety advantages. Some of them are designed to be built underground, protecting them from outside attack. Also, since they will generate less heat, some of the safety features of large reactors are not so necessary. Less water will also be needed for cooling.

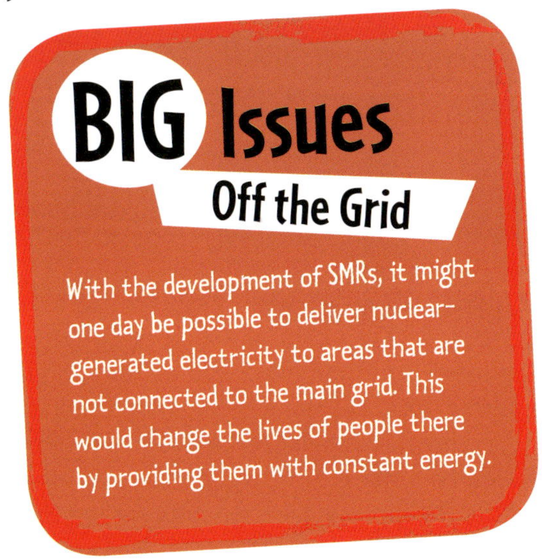

BIG Issues
Off the Grid

With the development of SMRs, it might one day be possible to deliver nuclear-generated electricity to areas that are not connected to the main grid. This would change the lives of people there by providing them with constant energy.

Working Together

The nuclear energy industry is large and complicated. For the industry to work well and move forward well, all the players in it—from governments and scientists to businesses—must work together to share their skills, knowledge, and experiences.

The Clean Energy Ministerial (CEM) brings together people from around the world to promote the change to **clean energy**, including nuclear power. These are some of the award winners at the 2014 Ministerial meeting.

Members on a Mission

The WNA represents the global nuclear industry. Its mission is to promote a wider understanding of nuclear energy among the main international bodies that have an influence over its development. Its members share their experience, and work together to improve their industry for the future.

Cleaning Things Up

Another organization bringing people together in this industry is the CEM. The group encourages people to look at all clean energy sources, including nuclear power. It has 24 countries as members (including the United States, Canada, the UK, Russia, Argentina, and Japan), who in 2018, launched Nuclear Innovation: Clean Energy Future (NICE Future). This is a program to help governments think about the role of nuclear energy alongside the other possible sources of clean energy, so they can work well together. It will also be looking at how nuclear technologies can be used in other energy industries and for other purposes.

In the United States, the Gateway for Accelerated Innovation in Nuclear (GAIN) and the Nuclear Innovation Alliance (NIA) are both working to support the development of nuclear technologies. It can be difficult for smaller-scale innovators to afford the setup costs for trials of their new ideas, so these organizations offer support and advice to help make this important work possible.

Taking Over the World!

It is important for the future of the nuclear energy industry that it has committed younger people working in it. When the current **generation** of experts **retires**, these younger people can take over. To help achieve this, in 2003, the World Nuclear University (WNU) was set up. It is a worldwide network for people in the industry or interested in joining it, and is committed to their training and education.

Nuclear industry workers from many countries meet to discuss safety in the industry, so nuclear power can play an important part in our future.

Nuclear Fusion

Nuclear fission is the process used in a nuclear reactor to produce the heat that creates steam to power the electricity generators. In this process, atoms of uranium-235 are split. There is another atomic process that also produces energy—and a lot of it. That is nuclear fusion. Fusion is when atoms are joined together to form a larger atom.

Star Power

Nuclear fusion is what powers the stars in the universe, such as our own sun. At the sun's core, or center, the incredibly high temperatures and the **force** of **gravity** cause the atoms of **hydrogen** there to be squeezed together. The hydrogen atoms become helium atoms and, in the process, masses of heat and light energy are transferred. In fact, enough is transferred to heat and light our whole planet, even though it is 93 million miles (150 million km) away!

If we could repeat that fusion process in a controlled way, we could generate enough energy to power the whole world forever. Unfortunately, we have not figured it out yet, but many people are working on it. It takes extremely high temperatures for fusion to happen, so the challenge is to find a way to create those temperatures, then maintain them long enough for fusion to happen. The process also needs very high pressure to squeeze the hydrogen atoms together.

If we can create "miniature suns" in a fusion reactor, we can generate unlimited amounts of energy.

This is the center of the tokamak fusion reactor (see below). This smart new reactor could help us unlock a huge amount of energy in the future.

Exciting New Reactors

There are several experimental fusion reactors around the world. The most promising is called a tokamak, which was originally designed in Russia. When hydrogen is heated to very high temperatures, its nuclei lose their electrons, and they both float free in what is called a **plasma**. The tokamak uses two forms of hydrogen called deuterium and tritium. Unlike uranium, they are not hard to find. Deuterium can be taken from seawater, and tritium can be produced in nuclear reactors from a common metal called lithium.

A tokamak aims to heat plasma to 180 million degrees Fahrenheit (100 million °C), seven times hotter than the center of the sun. The plasma is controlled at the core of the reactor using high-powered **magnets**. It uses an enormous amount of energy to reach these temperatures and pressure. Scientists have not yet reached the point of "energy gain," when they get more energy out than they put in.

BIG Issues
No Waste

Another huge advantage of nuclear fusion is that it produces only a small amount of radioactive waste. The main waste product is helium, a gas that we can reuse.

In the Long Run

The science behind the process of nuclear fusion has been proven and understood for years. The challenges of making it a usable source of energy are more practical; they involve materials and techniques. Around the world, many groups of scientists are working hard to make this a reality.

This is an aerial view of the International Thermonuclear Experimental Reactor (ITER), in Provence, France. ITER means "the way" in Latin, and scientists working at the site are trying to make nuclear fusion the way of the future.

Fusion Will Be the Future

The joke in the world of nuclear fusion has always been that it is forever 30 years away, partly because of the lack of government money to pay for its development. Now, however, several start-up companies have started working on fusion. One Canadian company is aiming to show that nuclear fusion can be a solution to our energy needs within the next five years. Another is working in the UK and is also hopeful of a breakthrough. They are working on a spherical tokamak, which they say works well. The spherical tokamak has achieved plasma temperatures of more than 27 million degrees Fahrenheit (15 million °C), and scientists are aiming for 180 million degrees Fahrenheit (100 million °C). They say that they expect to be supplying the grid by 2030.

Meanwhile, in the United States, the Massachusetts Institute of Technology (MIT) is working with a company to develop SPARC, a donut-shaped tokamak. The project has been partly funded by billionaires such as Bill Gates and Jeff Bezos, and it aims to develop reactors small enough to be built in factories and shipped for assembly on site.

Energy Supergiant

Internationally, 35 countries, such as the United States, the UK, China, India, Japan, and Russia, as well as countries in the EU, are involved in ITER. ITER is the world's largest tokamak, with 10 times amount of plasma of the largest tokamak operating today. Everything is on a huge scale. For example, nine suppliers have been working for more than seven years just to make the magnets needed to control the plasma. The first plasma is scheduled to be produced in 2025.

Experiment in Energy

ITER will not capture the energy it produces and use it to produce electricity. It is a research-and-development project, designed to show that energy gain can be achieved on a huge scale. Scientists will use it to study what happens to plasma, and to test technologies such as heating, control, and maintenance performed by robots.

Governments are funding the important research of ITER, so that it can lead to the development of fusion power plants by companies in the future.

CHAPTER 6

Nuclear Power Today

About 12 percent of the world's energy comes from nuclear energy today. In some countries, it provides more than half their electricity; in others, almost none. As we face the urgent need for clean energy generation, nuclear is up there on the list of technologies ready for new development.

Coming Back

Over the past 30 to 40 years, nuclear power slowed a little as an energy provider. Plants were very expensive to build and maintain, governments were not happy to spend a lot on them, and a lot of accidents made people uncertain about whether nuclear power was safe. Other sources of electricity, such as natural gas, also became cheaper. Today, however, that is changing. New designs for reactors are better, cheaper to build, and safer. Innovators are looking at every aspect of the industry. They are also looking for financial and practical support to get their new ideas off the ground.

New Countries Joining In

Nuclear energy is also spreading to other countries. Bangladesh, Belarus, Turkey, and the United Arab Emirates (UAE) are all constructing their first nuclear power plants. A number of other countries, such as Poland, Vietnam, and Egypt, are moving toward nuclear energy, too.

Turkey is relatively new to the nuclear power industry, but this thermal reactor is in operation in the country.

To protect our planet for future generations, experts believe that we need a combination of nuclear energy and other clean sources, such as wind and solar power.

Altogether, about 30 countries are planning or starting nuclear power programs, and 20 more are interested in joining them. Many less-developed countries are seeing a quick increase in the amount of energy people need as they become wealthier. This should come from clean sources, such as nuclear energy. Some of less-developed countries are working with countries with more-developed nuclear-power capacity, such as Russia and China.

New Designs

These places may be ideally suited to the newer SMRs that are being developed. These are less expensive to build and run, and they are also safer. SMRs are taking off in developed countries, too. In the United States, for example, the number of companies working with SMRs has increased by 56 percent in just three years. These are helping nuclear energy to be more competitive with other energy sources. As the old, larger nuclear reactors reach the end of their lives and have to be closed down, this new generation of plants will be ready to replace them.

BIG Issues

No Harm

If we are going to protect our planet, our energy future must be low carbon, meaning free of harmful emissions of carbon dioxide. Experts say that, at the moment, it is almost impossible for a major economy to have a reliable, low-carbon electricity supply without nuclear energy.

Evolutions of the Future

Nuclear energy seems to have an exciting future ahead. The technology is spreading to more countries, and some countries that have had it for a long time are setting aside some of their reservations about it.

A Brand New Fuel

In the future, fast breeder nuclear reactors may run on a different fuel altogether: a metal called thorium. Thorium is more common than uranium and also less radioactive. However, it cannot be used for fission directly. It is hit by neutrons fired at it by a device called a particle accelerator. The neutrons cause the thorium to turn into uranium.

Thorium is especially suitable for using in reactors that are cooled by helium gas rather than by water. The temperature in this type of reactor gets much higher, so more energy can be produced. Helium-cooled reactors are also cheaper than water-cooled reactors, because they do not need such an expensive cooling system. Thorium is also the ideal fuel for another new type of reactor of the future called a molten salt reactor, which is still at the design stage.

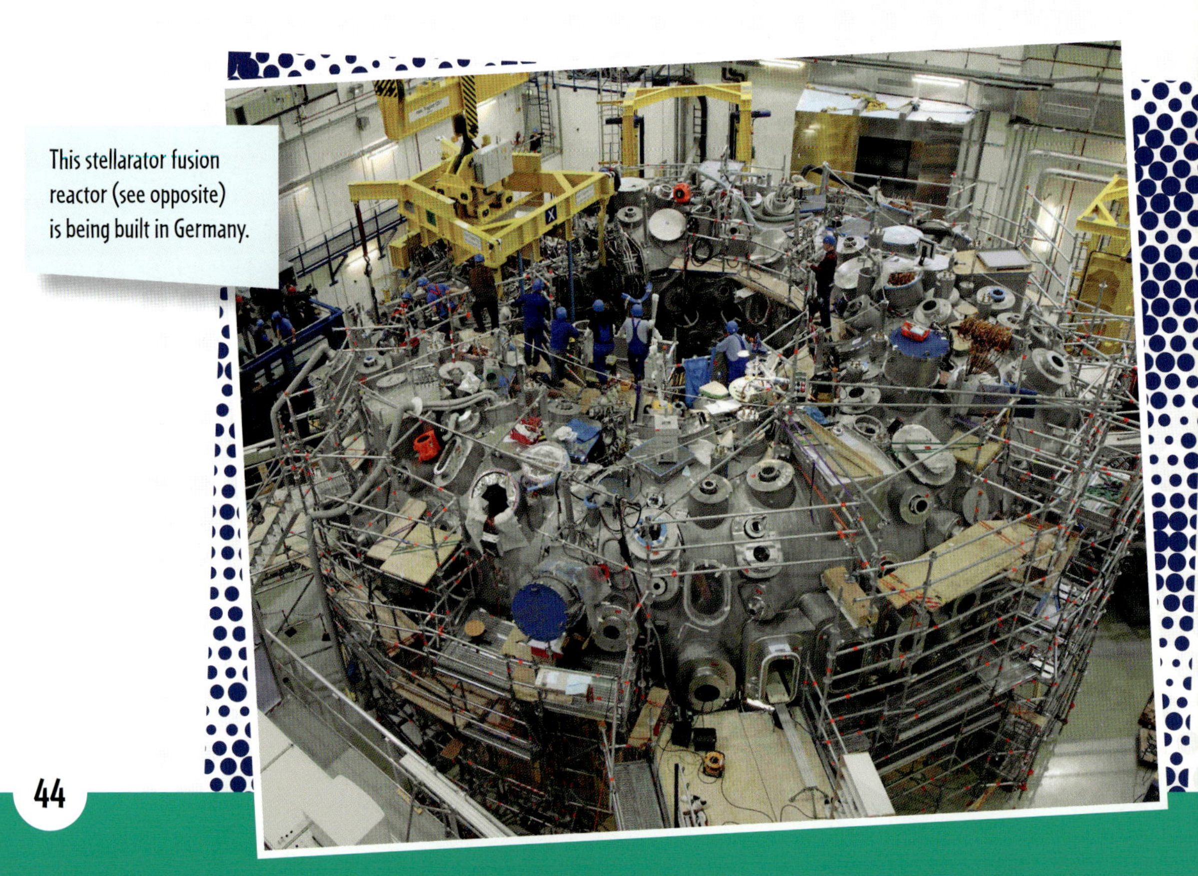

This stellarator fusion reactor (see opposite) is being built in Germany.

Help from Hydrogen

Another way in which nuclear power plants could be useful in the future is for making hydrogen from water. Hydrogen can be used in fuel cells to power cars, or it can be burned instead of gas to provide heat, without producing any harmful emissions. Vehicles powered by hydrogen fuel cells are already in use, and some people think we will need more hydrogen in the future.

Solving All Our Problems

The ultimate solution to all our energy needs is nuclear fusion. The fuel is plentiful, the reactors are safer to operate, and there is no harmful radioactive waste. Tokamaks are being tried out in several countries.

Another type of fusion reactor called a stellarator is also being trialed. This is able to hold the plasma in a more stable way than the tokamak, but it is more complex to design and build. The challenges of making fusion work are considerable, and we are not yet able to generate electricity that way. But when we figure it out, it will be the big one that changes our future.

One thing is certain, and that is that there is a rising demand for reliable, emission-free power around the world. If the nuclear energy sector can seize this opportunity, it should have a very bright future.

When we have plans to travel to Mars in the future, we will need to use nuclear power for the long journey to get us there. If we are going to try to live in that new world one day, we will need to send nuclear reactors separately on uncrewed cargo ships. They would be powered up only after we had left Earth, but they would give us the energy we would need so very far away.

Glossary

acid a powerful liquid that can burn, wear away, and break down soft materials and tissues
aluminum a common and lightweight silver metal
ambitiously with the intention of meeting high hopes
atmosphere the blanket of gases around Earth
budget the amount of money needed or available for a certain purpose
carbon dioxide a gas that gathers in the atmosphere and contributes to global warming
clean energy energy that does not harm the environment
climate change the changes in climate around the world caused by the gradual increase in the air temperature
competitive able to make great effort to gain or win over something else
coolants gases or liquids that remove heat from a reactor
depleted reduced in size, amount, or strength; used up
dissolved broken down in a liquid
efficiency the ability to do something well with little waste
efficient done well with little waste
emissions something, usually harmful, that is put into the air
environment the natural world
fertilizer material put into the soil to help plants grow
force strength or energy that can bring about change
fossil fuels energy sources in the ground, such as coal, oil, and gas, that are limited in quantity
generation all of the people born and living within a certain period of time
generator a machine that converts energy into electricity
geothermal energy energy that can be harnessed from the heat within Earth
global warming a gradual increase in the overall temperature of Earth
gravity a pulling force that holds objects on Earth
grid the network that distributes electricity from power plants to consumers
groundwater water that is found below the ground, not at the surface
helium a very light, colorless gas with no smell
hydrogen a very light, colorless gas that easily catches fire
innovation a smart new way of doing something
laser beams very powerful beams of light
lead a soft, silvery metal
magnets objects that can pull other magnetic objects toward them
nuclei the plural of nucleus; more than one nucleus
plasma a very hot gas in which the electrons are floating free among the nuclei
postponed arranged to take place at a later time than first scheduled
power plants places where energy is created
processed changed from its natural form
radioactive giving off radiation, which can have harmful effects
recycled changed and used again
remote far away and difficult to reach
renewable describes energy created from sources that do not run out, such as light from the sun, wind, water, and the heat within Earth
retires leaves ones job and stops working
samples small pieces or quantities taken from something, usually to study it
solar from the sun
solar panels panels designed to absorb the sun's rays for generating electricity
solar system the collection of eight planets and their moons that orbit around the sun
space probes uncrewed spacecraft that travel through space to collect scientific information
sustainable able to be used without being completely used up or damaged
tsunami a long, high, and very dangerous sea wave usually caused by an earthquake
turbines machines used to convert the movement of air or water into electricity

Find Out More

Books

Burgan, Michael. *Chernobyl Explosion: How a Deadly Nuclear Accident Frightened the World* (Captured Science History). Compass Point Books, 2018.

Honders, Christine. *Nuclear Power Plants: Harnessing the Power of Nuclear Energy* (Powered Up! A STEM Approach to Energy Sources). PowerKids Press, 2018.

Smith-Llera, Danielle. *Fukushima Disaster: How a Tsunami Unleashed Nuclear Destruction*. Compass Point Books, 2018.

Websites

Find out more about nuclear power at:
http://world-nuclear.org/nuclear-basics.aspx

The Energy Information Administration (EIA) takes an in-depth look at nuclear power at:
www.eia.gov/energyexplained/index.php?page=nuclear_home

Learn about the Nuclear Innovation Alliance (NIA) at:
www.nuclearinnovationalliance.org

Publisher's note to educators and parents:
All the websites featured above have been carefully reviewed to ensure that they are suitable for students. However, many websites change often, and we cannot guarantee that a site's future contents will continue to meet our high standards of educational value. Please be advised that students should be closely monitored whenever they access the Internet.

Index

About the Author

Robyn Hardyman has written hundreds of children's information books on just about every subject, including science, history, geography, and math. In writing this book she has learned even more about science and discovered that innovation is the key to our future.